FROM MUD TO MULCH: THE GREEN THUMB DIVA'S GUIDE TO GARDENING FUN

Nurturing the Next Generation of Growers with Fun, Easy-to-Follow Activities"

By: Evelyn T. Fitzgerald

COPYRIGHT PAGE

From Mic to Mulch: The Green Thumb Diva's Guide to Gardening Fun

Written by: Evelyn T. Fitzgerald AKA Shiba Shiba the Green Thumb Diva

Published by:
Stork Publishing LLC
www.lashawndashiree.info
334-232-9281
llove@lashawndashiree.info
Illustrations by:
Intellectual Designs by Lashawnda Love
Cover Design:
Intellectual Designs by Lashawnda Love
Copyright © 2024

All rights reserved. No part of this publication may be reproduced, distributed, or transmitted in any form or by any means, including photocopying, recording, or other electronic or mechanical methods, without the prior written permission of the publisher, except in the case of brief quotations embodied in critical reviews and certain other noncommercial uses permitted by copyright law. For permission requests, write to the publisher at the address above.

Printed in the United States of America

SHIBA SHIBA'S RADIO ADVENTURE

Hello, garden enthusiasts and little green thumbs! It's Shiba Shiba, the Green Thumb Diva, here to take you on a fun-filled journey into the world of gardening and nature!

I'm at the radio station, rocking my sparkling strawberry hat, with bright reds, greens, and a touch of glitter. Now, let's dive into this magical coloring activity book, packed with fun activities and colorful pages that make learning about nature a blast!
Grab your gardening gloves and crayons as we explore bees, plants, and how to care for our planet. Let's make this garden adventure unforgettable!

PLANT THE SEED RADIO SHOW

PLANT THE SEED. LET'S START GROWING!

Color the seed, soil, sun, & water-can! Follow the arrows to help the seed grow into a healthy plant.

COLOR THE SEEDS AND DRAW YOUR OWN!

Description: Draw and color different seeds. Label them with the type of plant they grow into. Can you find the sunflower seed, pumpkin seed, and apple seed?

Draw different seeds here

WELCOME TO THE ULTIMATE SEED MATCHING GAME EXTRAVAGANZA!

Hey there, my green-thumbed adventurers! 🌱 Shiba Shiba the Green Thumb Diva here, and I'm beyond excited to introduce you to an activity that's going to be so much fun, you'll be talking about it long after the seeds have sprouted! Get ready to dive into the world of seeds, plants, and imagination with the most epic Seed Matching Game you've ever played! 🎉

SEED MATCHING GAME: LET THE FUN BEGIN!

Step 1: Create Your Epic Seed Cards
- Materials Needed: Card stock, markers, crayons, colored pencils, scissors, and your boundless creativity!
- How to Create:
  - On one set of cards, draw or print pictures of different seeds—think sunflower seeds, pumpkin seeds, or even acorns! Make them as colorful and detailed as you can.
  - On another set of cards, draw or print pictures of the plants these seeds will grow into—towering sunflowers, bright orange pumpkins, or mighty oak trees! ♠
o Pro Tip: Use bright colors and add little fun details to make each card pop! Maybe your sunflower has sunglasses, or your pumpkin has a silly grin. Let your imagination run wild!

Step 2: How to Play the Seed Matching Game
- Get Ready: Shuffle the cards and lay them face down on a table or the floor. Make sure everyone has plenty of space to move around—this will be an action-packed game!
- Match Them Up: Take turns flipping over two cards at a time. Match the seed card with its corresponding plant card. When you make a match, shout, "Seed-tastic!" and strike your best gardener pose—because you're on fire!
- Discuss and Discover: After each match, take a moment to talk about the seed and the plant. How does this tiny seed grow into such a big, beautiful plant? What does it need to thrive—sunshine, water, a little TLC? The more you know, the more fun it gets!

Step 3: Level Up the Game!
- Team Play: Divide into teams and see who can make the most matches in a limited time. Set a timer and race against the clock! Tick-tock, green thumbs, tick-tock!
- Creative Twist: Want to spice it up? Try matching the seeds to a fun fact about the plant. For example, did you know sunflower seeds can follow the sun across the sky? Or that pumpkins can grow to be over 1,000 pounds?
- Seed Detective: Become a seed detective and try to guess the plant before flipping over the matching card. The fewer cards you flip, the more points you earn!

GROW YOUR FUTURE!

This isn't just a game, my friends—it's the first step into an exciting world of careers that help make our gardens and farms thrive!

- Agriculturalist: These plant pros study how plants grow and help farmers find the best ways to grow bigger, healthier crops. Imagine being the genius behind the juiciest tomatoes or the tallest cornfields!

- Seed Technician: If you love getting hands-on, this is for you! Seed technicians test and make sure seeds are top quality before they hit the fields. You're the gatekeeper of greatness—only the best seeds get your seal of approval!

- Botanical Illustrator: Got an eye for detail? Botanical illustrators create stunning, accurate drawings of plants and seeds that help scientists and gardeners alike. Your art could end up in textbooks or garden guides all over the world!

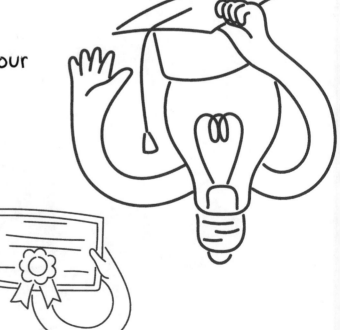

MEET THE PARTS OF A PLANT!

Color the plant & learn its parts!

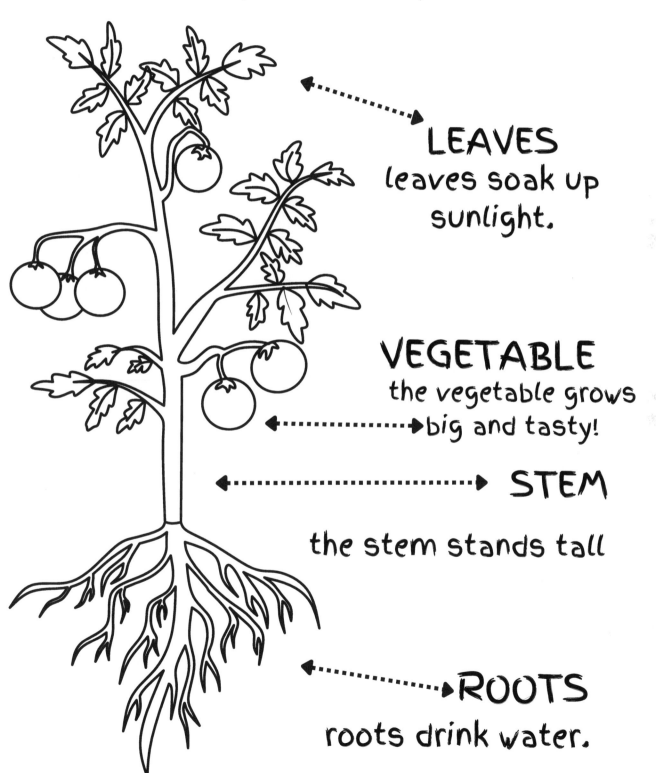

LEAVES
leaves soak up sunlight.

VEGETABLE
the vegetable grows big and tasty!

STEM
the stem stands tall

ROOTS
roots drink water.

PLANTS NEED FRIENDS TOO!

Match the plant friends. Instructions: Draw lines to match the plants with their pollinator friends like bees, butterflies, and birds. Color them in!

BEES

"Bees are excellent dancers; they use special moves to tell other bees where to find flowers."

BUTTERFLIES

"Butterflies can taste with their feet to find the best flowers for nectar."

ANTS

"Ants can lift objects that are 50 times their own weight, making them incredibly strong!"

LADYBUGS

Ladybugs are garden buddies because they love to snack on aphids (plant lice), which helps keep your plants strong and healthy!

BIRDS

"Some birds can fly thousands of miles without stopping for food or rest."

THE SUN IS A PLANT'S BEST FRIEND

Draw and color a big, bright sun in the sky over a garden. Can you draw plants reaching for the sun?

PLANT DETECTIVE ADVENTURE

Become a Plant Detective! Search your backyard or local park for different plants. Draw and color what you find in the boxes below. Can you find a flower, a leaf, a bug, and a tree?

Instructions:
1. Find a flower and draw it in the first box. What color is it?
2. Look for a leaf. Draw its shape in the second box. Is it smooth or bumpy?
3. Spot a bug near a plant. Draw it in the third box. What kind of bug is it?
4. Find a tree and draw it in the fourth box. Is it tall or short?

FLOWER

LEAF

BUG

TREE

PLANT YOUR IMAGINATION!

"Let's have some fun! Color this big pot or give it a happy face, then draw whatever plant or flower you want growing out of it. What will YOU grow today?"

VEGETABLES COME IN ALL SHAPES AND SIZES!

Vegetables come in all shapes and sizes! Some are round, some are long, and some are even funny-shaped. Color these veggies and see how different they are!

Cucumber

Apple

Radish

VEGGIES CAN BE FUN SHAPES

Create your veggie characters. Instructions: Draw faces on vegetables to turn them into characters. Color them in and give them names!

FUN GARDEN TIPS

PLANT A SEED:
DIG A SMALL HOLE, DROP IN A SEED, AND COVER IT WITH SOIL.

Watch it grow!

FUN GARDEN TIPS

WATER YOUR PLANTS: GIVE YOUR PLANTS A DRINK WHEN THE SOIL FEELS DRY.

They get thirsty, too!

FUN GARDEN TIPS

SUNNY SPOT: MAKE SURE YOUR PLANTS GET PLENTY OF SUNLIGHT.

Plants love the sun!

FUN GARDEN TIPS

HELP YOUR PLANTS GROW BY PULLING OUT WEEDS THAT STEAL THEIR FOOD AND SPACE.

FUN GARDEN TIPS

PLANT FRIENDS: GROW FLOWERS THAT ATTRACT BEES AND BUTTERFLIES.

They help your garden bloom!

FUN GARDEN TIPS

COMPOST MAGIC: USE KITCHEN SCRAPS LIKE VEGGIE PEELS TO MAKE COMPOST.

It's like plant food!

FUN GARDEN TIPS

COLORFUL GARDEN: PLANT FLOWERS OF DIFFERENT COLORS TO MAKE YOUR GARDEN BRIGHT AND BEAUTIFUL.

FUN GARDEN TIPS

GARDEN TOOLS: USE SMALL TOOLS LIKE A TROWEL OR WATERING CAN TO TAKE CARE OF YOUR PLANTS.

Welcome to "Plant the Seed Radio Show" with your host, Shiba Shiba the Green Thumb Diva! Today, we're diving into a fun and eco-friendly garden project that's perfect for all ages! We're making bird feeders from recycled plastic bottles, and it's going to be a blast. Let's get started!

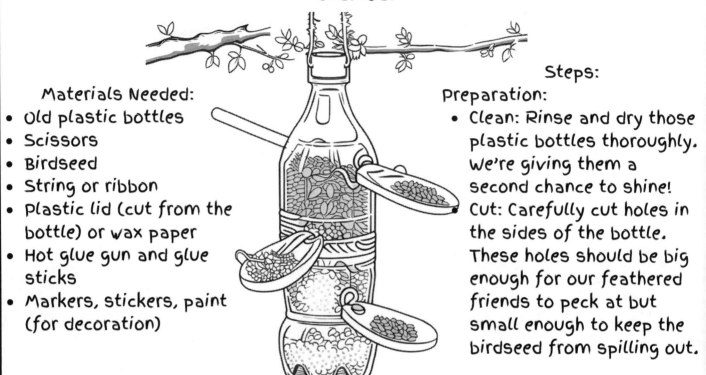

Materials Needed:
- Old plastic bottles
- Scissors
- Birdseed
- String or ribbon
- Plastic lid (cut from the bottle) or wax paper
- Hot glue gun and glue sticks
- Markers, stickers, paint (for decoration)

Steps:

Preparation:
- Clean: Rinse and dry those plastic bottles thoroughly. We're giving them a second chance to shine!
- Cut: Carefully cut holes in the sides of the bottle. These holes should be big enough for our feathered friends to peck at but small enough to keep the birdseed from spilling out.

Covering the Top:
- Cut the Cover: Use the top part of the bottle you cut off or a piece of wax paper to cover the top. If using the bottle's top, trim it to fit snugly over the opening.
- Attach: With your trusty hot glue gun, seal the cover onto the bottle's top. This will keep the rain out and ensure the seeds stay dry and delicious.

Decorate:
- Design: Let your creativity soar! Grab those markers, stickers, and paints, and deck out your feeder. The more vibrant and colorful, the better to attract our avian friends.

Add Feed:
- Fill: Pour birdseed into the bottle through the top. If you need a little help, use a funnel to make the job easier.

Hang:
- Prepare Hanging: Cut a piece of string or ribbon, thread it through the top of the bottle, and tie it securely.
- Hang: Place your feeder in a cozy spot where birds are likely to visit, like a tree branch or garden hook.

Introduction to livestock. (Rabbits & Quails)

1. **Rabbits:** are gentle animals that provide excellent fertilizer for gardens with droppings. They're fun to watch and help keep the garden healthy.
2. **Quails:** lay nutritious eggs that are a great addition to meals. They're small, easy to care for, and add variety to farm life.

Making Paper Rabbits or Quails
   Materials:
- Colored paper or construction paper
- Scissors
- Glue
- Markers or crayons

Steps:
1. Draw the Shape: On a piece of colored paper, outline a rabbit or quail. For a rabbit, draw large ears and a fluffy tail. Draw a rounded body and a small crest on the head for a quail.
2. Cut Out: Cut out the drawn shapes carefully.
3. Decorate: Use markers or crayons to add details like eyes, noses, and patterns. For rabbits, add whiskers and for quails, add feather details.
4. Assemble: Glue the pieces together if using multiple parts (like ears or tails for rabbits).

Display: Put the finished paper animals on display or use them in a farm-themed collage.

FUN FACTS FOR LIVESTOCK

RABBITS HELP IMPROVE SOIL HEALTH BY THEIR DROPPINGS,

which act as natural fertilizer.

RABBITS

"Rabbits help fertilize the soil naturally by their droppings, making them great for gardens."

FUN FACTS FOR LIVESTOCK

QUAIL EGGS ARE A NUTRIENT-DENSE FOOD,

packed with vitamins and minerals.

CHICKENS

Fun Fact: "Chickens aren't just for eggs! They love to munch on bugs, helping keep the garden free from pests. Plus, they're always ready to cluck a happy tune!"

COWS

Fun Fact: "Moo! Cows are amazing—did you know they help turn grass into rich milk? And their manure makes great fertilizer to help plants grow!"

GOATS

Fun Fact: "Goats are natural landscapers! They love to munch on weeds and help keep the farm clean. Plus, their milk is super nutritious and delicious!"

HORSE

"Horses have been helping humans in agriculture for thousands of years by pulling plows and carrying loads."

RADIO SHOW ACTIVITY FOR KIDS
SIMPLE AG-SHOW:

"The Amazing World of Farming!"

1. Preparation:
   - Provide kids with a toy microphone or a stick to use as a pretend mic.
   - Help them prepare a short "show" on a farming topic.
2. Subjects to Discuss:
   - "How Plants Grow": Talk about the lifecycle of a plant.
   - "Farming Tools Then and Now": Compare old tools to modern ones.
   - "The Animals on the Farm": Describe different farm animals and what they do.
   - "What Farmers Do Every Day": Share fun facts about a farmer's daily routine.

Sustainable Living Activity

"Eco-Friendly Craft Corner"

1. Project: Create items using recycled materials.
2. Materials: Old newspapers, plastic bottles, cardboard, and markers.
3. Steps:
   - Make bird feeders from plastic bottles or plant pots from cardboard.
   - Decorate and use them to help the environment.

Impact: Teaches kids about recycling and repurposing materials to reduce waste.

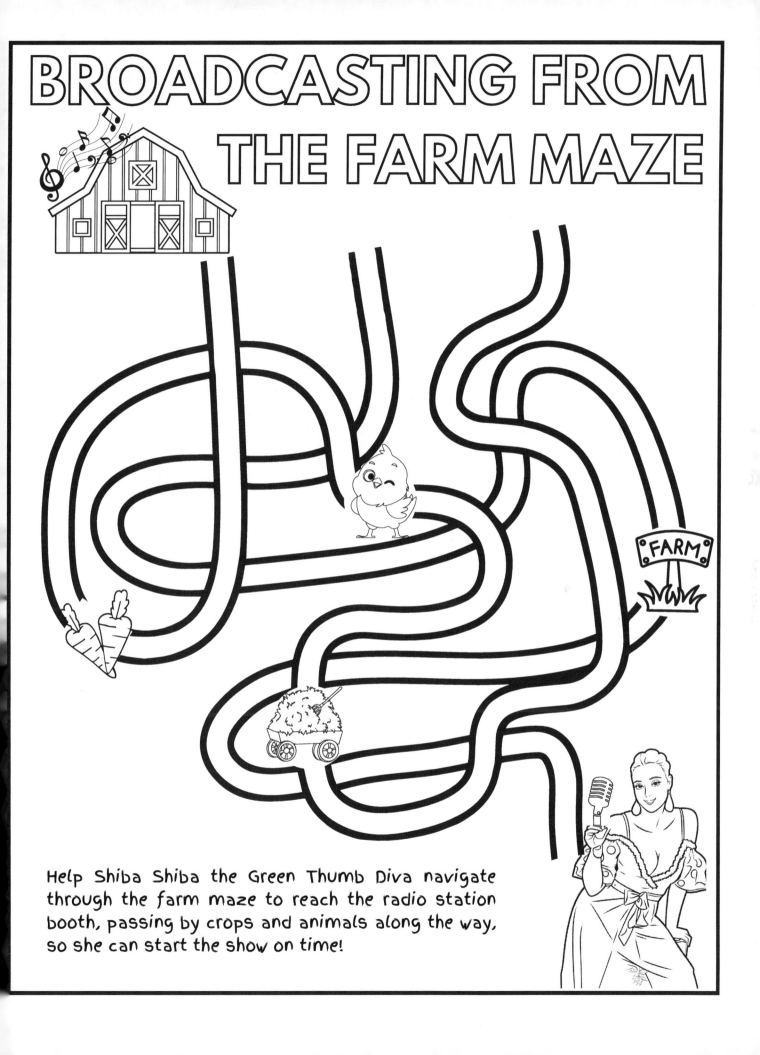

DESIGN YOUR OWN RADIO SHOW

Create your very own agriculture radio show! In the space below, draw or write about what your show will be like. Choose a fun name, decide what topics you'll cover—like gardening tips, animal care, or farming fun—and get creative with your show's theme!

CREATE YOUR OWN RADIO JINGLE ACTIVITY

"Plant those seeds, watch them grow,
In the garden, row by row!
Water, sun, and love each day,
Green Thumb Diva leads the way!
Gardens bloom, oh what a sight,
Growing strong and shining bright!"

Now it's your turn! Write and draw your own catchy radio jingle for a pretend agriculture show. Let your creativity bloom!

_____

_____

_____

_____

_____

_____

_____

_____

SHIBA SHIBA'S RADIO RECAP

It's time for your radio interview! Think about everything you've learned about agriculture in this book. Pretend you're being interviewed on a radio show—write down your answers to the questions below and share your thoughts on gardening, animals, and taking care of the earth!

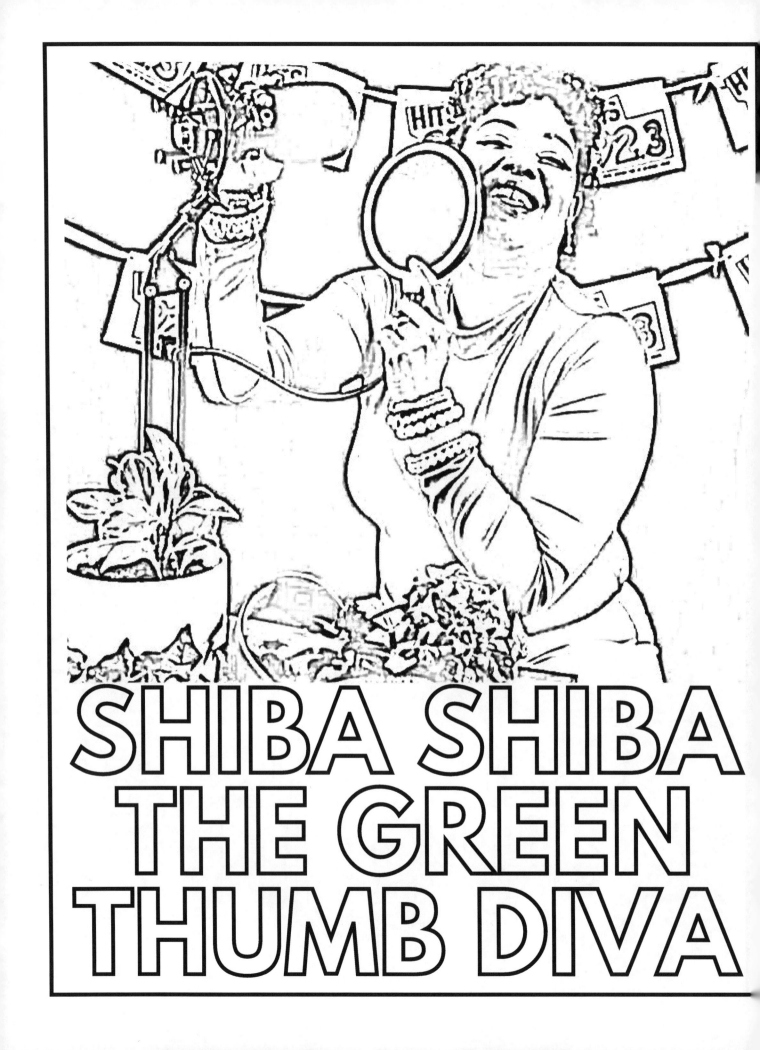

"WHAT'S A CORN'S FAVORITE MUSIC?"

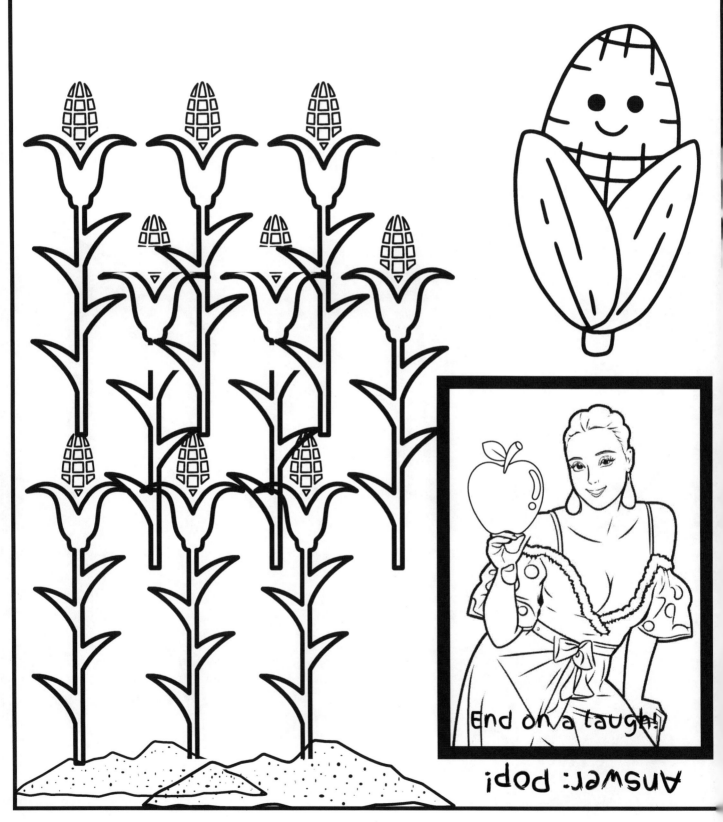

End on a laugh!

Answer: Pop!

Made in the USA
Columbia, SC
31 October 2024